Collins

easy learning

Numbers

Ages 3–5

Carol Medcalf

How to use this book

- Find a quiet, comfortable place to work, away from distractions.

- This book has been written in a logical order, so start at the first page and work your way through.

- Help with reading the instructions where necessary and ensure that your child understands what to do.

- This book is a gentle introduction to written numbers, built up in order, with clear, easy-to-see corresponding quantities. Try to use the following language as you work through the book together: numbers 1 to 10, next, first, last, more, less.

- If an activity is too difficult for your child then do more of our suggested practical activities (see Activity note) and return to the page when you know that they're likely to achieve it.

- Always end each activity before your child gets tired so that they will be eager to return next time.

- Help and encourage your child to check their own answers as they complete each activity.

- Let your child return to their favourite pages once they have been completed. Talk about the activities they enjoyed and what they have learnt.

Special features of this book:

- **Activity note:** situated at the bottom of every left-hand page, this suggests further activities and encourages discussion about what your child has learnt.

- **Number train:** situated at the bottom of every right-hand page, this highlights in white type the number that is being learnt and also shows the numbers that have been covered in black type.

- **Counting activity:** spend time on the first activity on every double page. Point at the numeral on the child's T-shirt in the picture. Show your child how to count the number on their hand. Look closely at the red counter/s and help your child to count them.

- **Trace the number:** talk about the shape of each number and the pencil movement as they write – straight down for number one. Is the number significant to your child? Their age or house number perhaps? This will help your child to remember specific numbers more easily.

- **Certificate:** the certificate on page 24 should be used to reward your child for their effort and achievement. Remember to give them plenty of praise and encouragement, regardless of how they do.

Published by Collins
An imprint of HarperCollins*Publishers* Ltd
The News Building
1 London Bridge Street
London
SE1 9GF

Browse the complete Collins catalogue at
www.collins.co.uk

© HarperCollins*Publishers* Ltd 2006
This edition © HarperCollins*Publishers* Ltd 2015

10 9 8 7 6 5 4 3

ISBN 978-0-00-815154-6

The author asserts the moral right to be identified as the author of this work.

All rights reserved. No part of this publication may be reproduced, stored in a retrieval system, or transmitted, in any form or by any means, electronic, mechanical, photocopying, recording or otherwise, without the prior permission of Collins.

British Library Cataloguing in Publication Data

A Catalogue record for this publication is available from the British Library.

Written by Carol Medcalf
Design and layout by Lodestone Publishing Limited and Contentra Technologies Ltd
Illustrated by Jenny Tulip
Cover design by Sarah Duxbury and Paul Oates
Project managed by Sonia Dawkins

MIX
Paper from responsible sources
FSC™ C007454

Contents

1 One

- Count 1.

- Trace the number 1.

- Colour 1 apple.

To help your child form the shape of numbers before writing them, cut numbers out of sandpaper and stick them onto card. Help your child trace their finger over the sandpaper to feel and learn the number shapes.

- ## Count 1. Write the number.

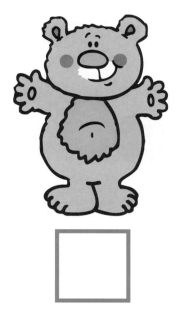

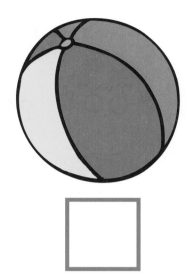

- ## Draw a circle round the odd one out.

2 Two

- ## Count 2.

- ## Trace the number 2.

- ## Pairs come in twos. Draw a (circle) round the matching pairs.

- Which bowl has two fish? (✔)

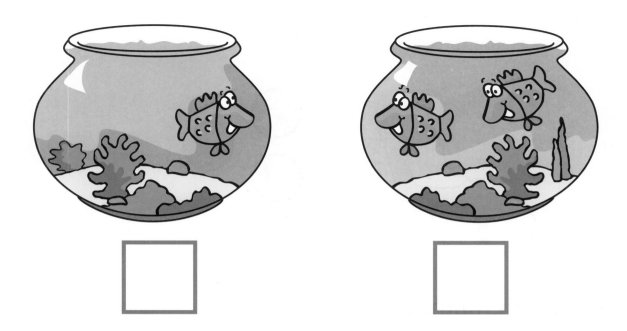

- Follow the path to take the twins to house number 2.

3 Three

- **Count 3.**

- **Trace the number 3.**

3 3 3 3 3

- **Draw a** (circle) **round kite number 3.**

Look closely at the grouping of red counters for each number. Help your child to understand the patterns; this will help when they start addition.

● Count the bears. Write the number.

● Colour three bowls of porridge.

● Draw a (circle) round the set with 3 chairs.

4 Four

- Count 4.

- Trace the number 4.

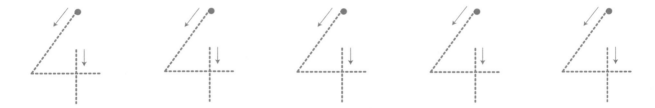

- How many sides has a square got? Draw a (circle) round the correct number.

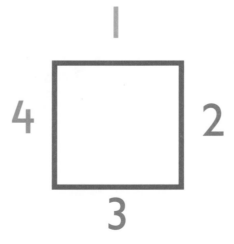

Try to provide your child with lots of counting experiences. Cut out 10 squares of card and write one number on each one (1 to 10). Using counters, help your child match the right number of counters to each card.

● Draw a circle round the box of four cakes.

● Colour the number fours. What number can you see?

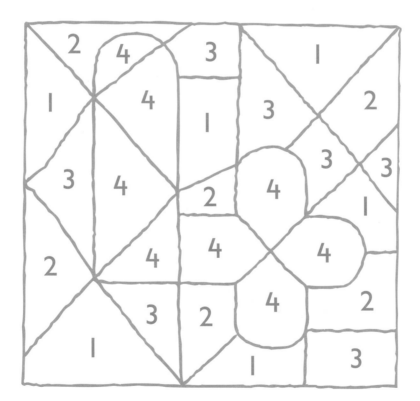

5 Five

- Count 5.

- Trace the number 5.

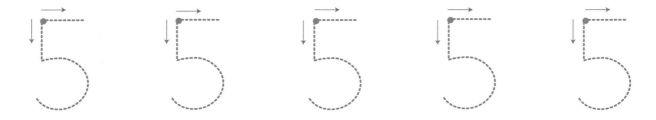

- Count the fingers. Write the number.

When you're out together, count everything that you see, such as vehicles, animals, buildings, etc. Make a chart and record the number of each thing that you see.

● Draw a (circle) round the plate that has 5 sandwiches.

● Find all the fives in the picture.

6 Six

- Count 6.

- Trace the number 6.

6 6 6 6 6

- How many spots has the ladybird got? Draw a circle round the correct number.

4

5

6

1

2

3

Use mathematical language when playing with toys: 'You have two, can I have one?'/
'Let's build towers that are nine bricks high.'

Which set has six sweets? (✔)

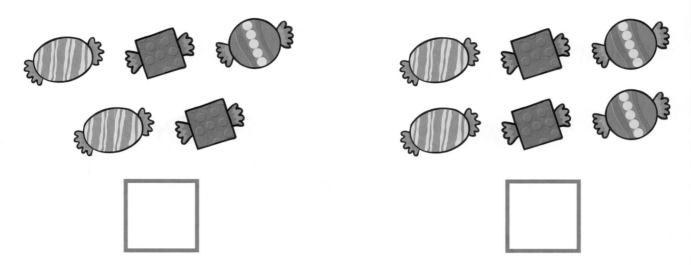

Trace the route for bus number six.

7 Seven

- Count 7.

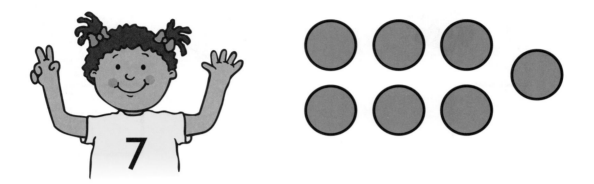

- Trace the number 7.

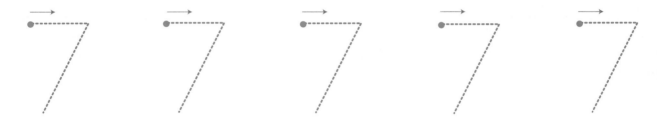

- How many butterflies are there? Write the number.

Collect small boxes of various shapes and sizes. Put some small toys in each box and take it in turns to shake them and guess how many toys are in each box. Count them out together and see who was correct. Introduce concepts such as 'more' and 'less'.

● Colour seven stars.

● Draw seven candles on the cake.

8 Eight

- Count 8.

- Trace the number 8.

- How many legs has the octopus got?
Write the number.

Sing number songs/say number rhymes together: 5 little ducks, 10 green bottles, etc.
Can you think of one for each number?

The farmer has lost his sheep. How many can you find? Write the number.

9 Nine

- Count 9.

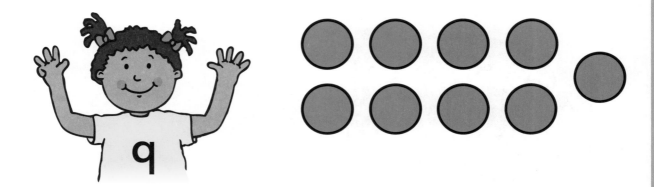

- Trace the number 9.

- Colour 9 balloons.

Count out seeds and plant them together. Write the number of seeds planted on the flower pot and then see how many grow! This can be used to develop basic counting skills, ask: 'How many of the seeds that you planted, grew?'

● Colour fish number nine.

● Follow the numbers in order. Can you find the cheese?

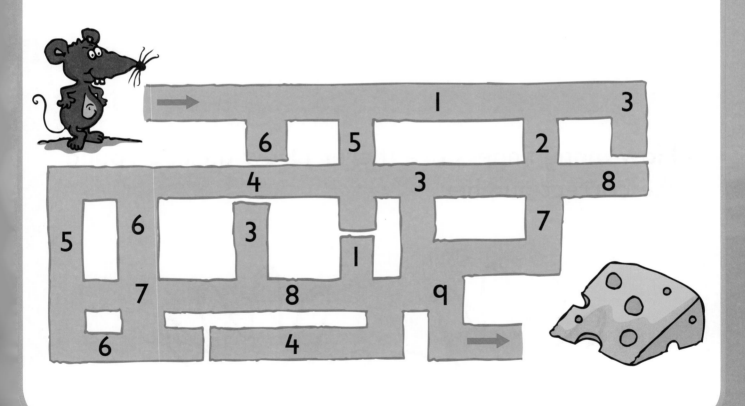

10 Ten

- ## Count 10.

- ## Trace the number 10.

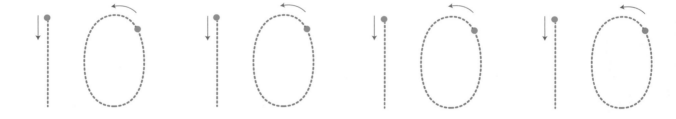

- ## How many fingers are there? Draw a circle round the correct number.

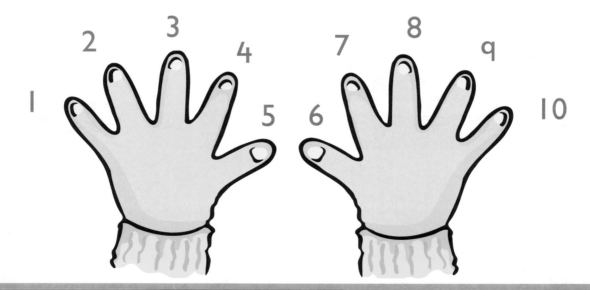

Using the number cards made on page 10, get 10 pegs and see if your child can peg the numbers on a washing line in the correct order.

● Which jar has ten sweets? (✔)

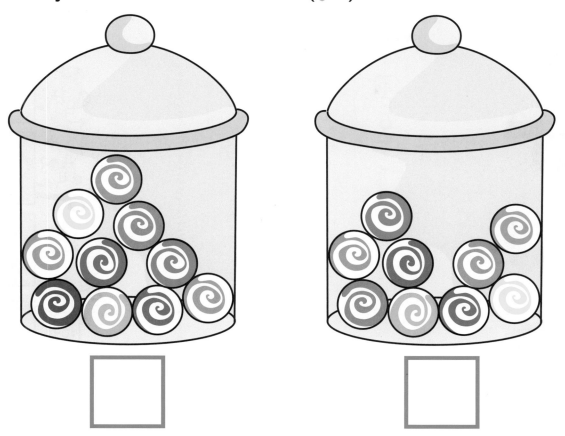

● Colour ten cherries on the cake.

1 2 3 4 5 6 7 8 9 10

Well done .... (name)

You have finished!

Now you know the numbers 1 to 10!

Age ....

Date ....